A
Rule
Concordance
and
Value Guide

Philip E. Stanley

2004 Edition

Astragal Press

First Globe Pequot edition 2019

Published by Astragal Press
An imprint of Globe Pequot Press
Wholly Owned by: The Rowman & Littlefield Publishing Group, Inc.
4501 Forbes Boulevard, Suite 200
Lanham, Maryland 20706

Distributed by National Book Network
1-800-462-6420

A Rule Concordance and Value Guide

This rule concordance and value guide presents an overview of the number and variety of rules manufactured by the major American rule makers during the years 1840-1940, arranged in tabular form to allow direct comparison of rule numbers and to illustrate the wide diversity of rule patterns available during this period. For each type of rule an estimate is given of the price which that rule would sell for on the open market, this price being expressed as a range from the price of a rule in GOOD condition to one in NEW condition.

TABLE STRUCTURE AND RULE NUMBERS

Beginning on page ten is a series of tables, each concerned with a different basic type of rule ("1 FOOT, 4 FOLD WOOD RULES," "2 FOOT, 2 FOLD IVORY RULES," etc.). Each table consists of a grid of rows and columns, where the rows represent the different patterns (e.g., joint, trim, width) found in the basic type of rule covered by that particular table, and the columns represent the different makers (and for some makers different periods of manufacture). Thus a box at the intersection of a row and column in a table represents a particular pattern of that basic type of rule by a particular maker. An extra column on the left hand side of each table has entries describing the particular pattern associated with each row, and a space is provided to the right of the rightmost column for comments to identify distinguishing features of rules which are otherwise identical. It follows, then, that any given rule can be found in the intersection of a row and column in some one of the tables, that table and intersection identifying the maker and type/pattern.

The presence of a number "**NN**" in any box indicates that a rule of that type/pattern was offered at some time by that maker under that number. "**(No #)**" in a box indicates that that maker offered a rule of this type/pattern in a catalog without assigning a number to it, "**???**" that an unnumbered rule of this type/pattern marked with that maker's name has been observed.

When a maker has offered several different rules of the same type/pattern, a separate row is provided for each rule, with the particular feature(s) which distinguish that rule from the others identified by notes to the right of the rightmost column.

Rules listed in the same row in different columns are essentially identical as to their mechanical construction and graduation. Rules listed as being the same type/pattern, but in different rows, are essentially identical as to their mechanical construction, but differ significantly in some other feature (e.g.: The Stanley No. 82 had board tables, as did the Stearns No. 14; the Stanley No. 76 did not).

Rules are classified into Narrow, (Regular), and Broad as a function of rule width when folded. For rules folding to three inches, regular width is considered to be $5/8$ inch; for rules folding to four, six, or nine inches: 1 inch to $1 1/8$ inches; for rules folding to one foot: $1 3/8$ inches to $1 1/2$ inches.

Unless otherwise noted, wood rules have yellow brass joints and trim; ivory rules have German silver (white brass) joints and trim.

The information regarding rule patterns and numbers upon which this concordance is based has been derived from manufacturers catalogs and price lists, wholesalers lists of comparative rule numbers, and from observed examples. The lists of comparative rule numbers were particularly useful in the case of Stephens rules, where only two known price lists exist, and the Standard Rule Co., where no price list has ever been found.

MANUFACTURER-SPECIFIC COMMENTS

Only the major makers have been listed, makers for whom there is adequate information in the form of examples, catalogs and price lists to characterize their product lines. These makers are:

> A. Stanley & Co./The Stanley Rule & Level Co.
> E. A. Stearns & Co.
> H. Chapin
> Stephens & Co.
> The Chapin-Stephens Co.
> Belcher Brothers & Co.
> Standard Rule Co./Upson Nut Co.
> Lufkin Rule Co.

The many smaller rule makers, A. Gifford, J. Watts, Clark & Co., for example, have been omitted; there is not enough information about their lines of rules to allow their being properly placed in the tables. Also omitted were the specialty rule makers such as Kerby & Brothers and Merrifield & Co. Kerby was a major maker, but its product line was composed almost exclusively of specialty measuring instruments which are outside the scope of the tables; Merrifield seems to have specialized almost exclusively in the manufacture of Gunter's rules, and did not have a broad enough product line to justify inclusion.

Stanley Rules

The rules manufactured by A. Stanley & Co. up to 1857 have been listed in a separate column from the rules made by The Stanley Rule & Level Co. in 1859 and thereafter. Rules made by A. Stanley & Co. are highly valued by collectors, and command much higher prices than the products of the successor firm; it was necessary to list them separately in order to properly provide reasonable value estimates for both product lines.

Stearns Rules

The line of rules offered by E. A. Stearns & Co. before 1856, when E. A. Stearns himself was operating the company, differed significantly from the line offered after 1856, under the proprietorship of Charles L. Mead, and later the Stanley Rule and Level Co. The two lines differed both in the patterns of rules offered, and the system of numbers assigned to the rules. These two lines/systems are listed separately in this concordance in two adjacent columns labeled Stearns (early) and Stearns (late).

Assignments of numbers in the Stearns (early) column marked "(*)" are tentative. The only known price list from before 1856 is so cryptic in its descriptions as to make it difficult to place these rules in the table with certainty.

Chapin Rules

Sometime between 1853 and 1859 Hermon Chapin made some changes in his line of rules and at the same time adopted a totally new rule numbering system (his son, Edward, had joined the firm at about this time, and this may have been due to his influence). These two lines/systems are listed separately in this concordance in two adjacent columns: Chapin (early) and Chapin (late). The Chapin (early) rules were not marked with the rule number, but only with Hermon's name and the phrase UNION FACTORY; apparently the numbers were only used for catalog and ordering purposes. The Chapin (late) rules were marked with the rule number.

Belcher Brothers Rules

It is doubtful if this company actually manufactured and stocked all the minor variations of rule pattern which were listed in its catalog and are shown in this concordance. What is more likely is that it listed all these combinations of length, joint, trim, width, etc. in order to enhance its prestige, but that many of them were not stock items, and were never made except on special order. Like Chapin (early), Belcher Brothers only used numbers for catalog and ordering purposes, and never marked them on the rules themselves.

Lufkin Rules

The Lufkin tailors' and glaziers' rules and squares have been omitted from this concordance as being unique to Lufkin, and thus not comparable to the products of any other major maker. The Lufkin board and log sticks have also been omitted; other major makers offered implements of this class, but the Lufkin sticks were so numerous and so unique in their construction as to also constitute a separate line.

VALUE ESTIMATES

In each box where a rule number is shown, thus indicating the existence of a rule of that type/pattern by that maker, an estimated value is given for that rule, in the form "$NNN-$NNN." This estimate represents the price which this type/pattern of rule by this maker could be expected to bring when sold on the open market. The estimate is stated in the form of a range, because the price will depend heavily on the condition of the rule, and also on the selling environment and other factors (see the discussion below).

For estimating purposes rules are divided into seven condition categories: NEW, FINE, GOOD+, GOOD, GOOD-, FAIR, and POOR. The upper figure in a given value range represents a reasonable estimate for that rule in NEW condition; the lower figure, a reasonable estimate for that rule in GOOD condition. Rules in GOOD+ and FINE condition will have values between these extremes, with FINE being about midway between GOOD and NEW, and GOOD+ about midway between GOOD and FINE. Rules in GOOD- and FAIR condition will have values of about 1/2 and 1/4, respectively, of the GOOD value. Rules in POOR condition are almost worthless.

If this value scale seems a little skewed, with only the finest rules in the upper end of the range, remember that condition is *extremely* important to rule collectors. They prize original finish and unworn condition (lacquer still on the brass, sharp edges, ivory not yellowed, etc.) *very* highly, and rapidly lose interest if these conditions are compromised (see the section CONDITION, below).

The value estimates in this concordance are based on selling prices observed during more than five years of research at auctions, in dealer catalogs, at dealer sales, on the internet, and through discussions with knowledgeable rule collectors. Only actual selling prices have been given weight; asking prices only represent what the seller hopes to get, not what someone is willing to pay for it, and the latter amount, after all, is what something is worth.

RELIABILITY OF ESTIMATES

The reliability of these estimates depends significantly on the availability of hard sales data. A maker whose rules are widely traded provides extensive data upon which to base value estimates for his rules; a maker whose rules rarely find their way to market provides little or none.

The most commonly traded rules are those of the Stanley Rule & Level Co. Many hundreds of Stanley rules are sold every year, providing an abundance of hard sales data. Additionally, there are a very large number of Stanley tool collectors, and as a result Stanley tools are the subject of intensive study, and documentation, including, in particular, John Walter's landmark *Antique and Collectible Stanley Tools*. Value estimates for these rules, based on Walter and on recent selling prices, can be taken as reliable and timely.

The maker whose rules appear least often on the market is Belcher Brothers. Except for their Gunter's scales and shrinkage rules, Belcher Brothers rules are almost never seen, and value estimates for their rules are perforce based primarily on collector and dealer opinions.

In terms of scarcity, the rules of the other major makers fall between those of Stanley and Belcher Brothers as shown in the list below, in the order shown.

Most Common	Stanley Rule & Level
	Lufkin Rule
	Upson Nut
	Chapin-Stephens
	Chapin
	Stephens
	E. A. Stearns
	A. Stanley
	Standard Rule
Scarcest	Belcher Bros.

Estimates for rules from these other major makers are derived from a combination of reported sales data combined with collector and dealer opinion, as appropriate.

FACTORS AFFECTING VALUE

A number of factors affect the value of a measuring instrument to a collector, the two most important being condition and scarcity. It is hard to say which of the two is the most important, but it is unquestionably true that either one has a larger effect on value than any other factor such as material and finish, selling venue, or charisma. This is not to say that these last three are not important; only that their influence on value is significantly less than the effect of either of the first two.

Condition

As mentioned earlier, condition is *extremely* important to rule collectors. They prize original finish and unworn condition (lacquer still on the brass, sharp edges, ivory not yellowed, surface finish intact) *very* highly, and rapidly lose interest if these conditions are compromised. Stains, loose joints, wear to the finish or metal trim, or other signs of wear can put them off completely. It is easy to see why this is so. More so than most tools, the appeal of a measuring instrument is in its appearance: the elegant brass or German Silver trim, the crisp precise graduations, the fine lacquer or shellac finish on virtually grainless boxwood. Wear, damage and staining all destroy that fine appearance, and with it much of the rules' value to a collector.

Scarcity

As a general rule, the scarcer something is, relative to demand, the higher the price it will command in the marketplace. In economics this is called "the law of supply and demand," and it is as true for antique rules as it is for any other merchandise.

As an example, consider the Stanley Nos. 67 and 68 two foot, four fold rules, identical except for width. The No. 68, 1 inch wide, was Stanley's basic, least expensive rule, and was offered well into the 1950s; nice examples are commonly encountered even today, and can be acquired for only about $15.00. The No. 67, 1⅜ inch wide, was only offered until 1917; examples in any condition are correspondingly scarce today, and a clean one will cost a minimum of $100.00 to $200.00.

Keep in mind, however, that *demand* is also an important factor in the value of a rule. Every rule collector would love to own a Stephens No. 38 ivory combination rule; they are both visually spectacular and a "holy grail" of rules. However, the number of collectors who have the thousands of dollars currently asked for these rules is severely limited (what the economists call a "shallow" market). If two dozen of them were suddenly to appear on the market many of them would go begging at the current high price.

There is one reverse effect in the relationship between supply and demand as it applies to collecting that is worth noting. Collectors like to feel that the item they are collecting can be found, given sufficient diligence, persistence, and patience (and money); that they can continually build and expand their collection. When something is so scarce that it is almost impossible to find new examples, they lose interest in collecting that thing, and gradually drift into collecting something else. A good example of this are the rules made by Belcher Brothers & Co. As mentioned earlier, its four fold rules are the least frequently found of any major maker. This scarcity should make them more desirable than those of any of the other makers covered in this concordance, but that is not the case. This extreme scarcity makes it almost impossible to build a good-sized collection, and as a result collectors do not pursue them, and their prices are not as high as would otherwise be the case.

Material & Finish

The material that a rule is made of also influences its value. A crisp boxwood rule with a nice patina is beautiful, but to a collector the smooth, white appearance of ivory is much more desirable. Exotic woods, such as rosewood or ebony, are also valued highly. These woods were only occasionally used for rules; they were not as fine-grained as ivory and boxwood, and their darker color made it difficult to see and read the graduations. When they were used, the graduations lines and figures were filled with a white, instead of a black, pigment to improve visibility. Modern rules, often made of maple or some other lesser wood, and sometimes with a painted surface, are not highly valued.

Similarly, among joint and trim materials German silver is more desirable than brass. Though basically just brass with a 20% admixture of nickel, its white, silver-like appearance is visually very satisfying and makes rules so trimmed much in demand.

Another factor is the level of finish of a rule. Some early makers (L. C. Stephens was one) actually polished their rules after finishing them, giving them a beautiful, smooth surface. Others simply left the lacquer or shellac as deposited by the brush, and the slightly rough resulting surface is detectable. To many collectors, the first condition is preferable to the second, and has a significant influence on value.

Selling Venue

Where and *how* a rule is sold can significantly influence the selling price. The closer you are to the original seller, the less that seller (presumably) knows about the nature of the rule and what its true value is, and the less he or she will likely ask for it. The closer you get to the final collector, knowledgeable and determined, the higher the price becomes.

The lowest prices are probably those asked by sellers to whom a rule is just one more in a bunch of tools to be disposed of. Such sellers are most often found at local yard/garage sales, and at flea markets, particularly those small, local flea markets which are only held once a year, and which have a high percentage of homeowners, as compared to professional flea market dealers.

Higher prices usually prevail at venues where the seller has some knowledge of rules. There are a large number of pickers and general antique dealers in the trade who own a small reference library and have some general knowledge of tools and measuring instruments. They are not specialists in tools, but they know enough to not sell a good rule for a "handful of beans."

The highest prices are probably those paid at major tool auctions such as those held semiannually in Nashua, New Hampshire, Harrisburg, Pennsylvania, and in England. Major collectors with ample resources attend these well-cataloged auctions in pursuit of the few exceptional rules offered, and the competition is intense (the phrase "feeding frenzy" is sometimes used to describe the resulting frantic bidding).

A rule can pass through several hands, at ever-increasing prices, before it finally arrives in a collection or at a major tool auction. These intermediate stages are "pickers," general antiques dealers, and a specialist antique tool dealers.

A "picker" is a source-oriented intermediate antique dealer. He goes to local yard sales and flea markets, and maintains contacts with a network of house content dealers, buying items cheaply which he recognizes as worth more than the seller is asking. He does not have a store or a stall in a group shop, but disposes of what he buys in private sales and auctions. He is not usually connected to the general body of collectors, but sells rather to one or more dealers and auctioneers, or (more and more commonly) in internet auctions. Some pickers regularly sell in the parking lot at tool collector meetings and tool auctions. A picker hopes to get a collector price for a rule that he is offering for sale, but will usually settle for significantly less; typically around 50% (but this varies widely). Pickers are worth cultivating as an occasional, if irregular, source of good rules. Chat them up, and try to develop a good relationship with them. Don't nickle and dime them by paying a pittance, even if they don't know enough to ask anywhere near the true value; if you do, and they find out, they will be reluctant to deal with you in the future.

The next step up the chain is the general antique dealer. These dealers are destination-oriented intermediate owners, selling rules either directly to collectors or consigning them to auctions (usually tool auctions). They sell in shops (either their own or group shops), antique shows, and at major antique flea markets (such as Brimfield). Many general antique dealers will sell rules for what is really a wholesale price, either through ignorance, or in the interests of a quick turnover. Others, however, will seek a collector price (or more).

Antique dealers who specialize in tools ask even higher prices. They are knowledgeable sellers, and often collectors in their own right. They know the market, and which pieces are most desirable, and usually set their prices at or just below what they consider a good collector price. They are a good source of rules for your collection, but be prepared to pay a collector price for what you get.

Charisma

Certain rules, or the rules of certain makers, are perceived by collectors as more desirable than other rules of equal quality and complexity, and command correspondingly higher prices. This property is frequently referred to as the "charisma" of a rule.

Charisma is not a function of scarcity. If it was, the rules made by the Upson Nut Company would command higher prices than those made by Stanley. Stanley rules have it; so do rules made by Stearns and (to a certain extent) Belcher Bros. and Standard Rule. Upson Nut and Lufkin rules do not. The other makers, Chapin, Stephens, and Chapin-Stephens lie somewhere in between.

Neither is charisma purely a function of maker. Certain patterns or particular models of rules are perceived as desirable by collectors, and will command correspondingly higher prices than other equally scarce rules by the same maker. A good example of this is the No. 36 Stevens patent combination rule, made for almost 100 years by (successively) Stephens, Chapin-Stephens, and Stanley. A Stephens No. 36 in GOOD condition will command easily three or four times the price of any other Stephens four fold rule, even one with elaborate scales or tables, in similar condition. Collectors just like the No. 36, and will pay a premium price for it.

CONDITION

As indicate earlier, the condition of a rule is almost as important as maker and pattern in determining its value. The scale generally accepted in the tool collecting community for grading tool condition was originally developed in about 1980 by Vernon Ward for use in describing tools in his *Iron Horse Antiques* tool catalog, and is still used today, in an improved form, by Clarence Blanchard in the *Fine Tool Journal* and John Walter in *Antique and Collectible Stanley Tools*. This scale grades tools as either NEW, FINE, GOOD+, GOOD, GOOD-, FAIR, or POOR, and when used is accompanied with a table equating each aspect of a tool with each grade.

This concordance uses the same scale (NEW, FINE, GOOD+, GOOD, GOOD-, FAIR, and POOR), creating a modified table (below) which focuses specifically on the features of rules and measuring instruments.

RULE CONDITION CHART

RATING	Surface Condition [% Original Finish]	Edges, Corners, & Joints	Sticks	Graduations	Problems
NEW	Flawless, probably unused [95%-100%]	Edges/corners sharp, joints tight	Straight	Sharp and black	None
FINE	Near perfect, barely visible wear [85%-95%]	Very slight edge and corner wear	Straight	Sharp and black	None
GOOD+	Slight wear [70%-85%]	Some wear, but still almost crisp	Very slight warping	Worn at edges	No major problems
GOOD	Some wear [50%-70%]	Significant edge and corner wear	Slight warping	Worn but readable	No major problems
GOOD-	Significant wear [35%-50%]	Well used, joints loose	Significant warping	Badly worn	Scratched, chipped, barely usable
FAIR	Almost worn off, w/stains/dirt [20%-35%]	Much wear, joints floppy	Warped, bent, split	Barely readable	Unusable
POOR	Worn off, w/bad stains/dirt [less than 20%]	Severe edge wear, joints broken	Warped, bent, broken	Worn away	Unusable

SOME GENERAL NOTES ON CONDITION

Original Box

Measuring instruments are rarely encountered still in their original box. Manufacturers usually packaged rules in accordion boxes containing dozen, 1/2 dozen or a dozen rules, and that is how they were stocked in stores. A craftsman buying a rule would not receive it in the box, as he would if buying some larger tool which was packaged singly. The box would remain in the store to hold the remaining rules, and the buyer would receive only the rule.

Some premium rules were packaged one to a box. The 36/036 Stephens patent combination rules were so packaged, as were the Long's patent gear rules and Hogg's patent textile rules. So were rules which were modified with special features by second parties, such as the Marsh dry measure gauge. These were exceptions, however, and not the general rule.

Because the box remained in the store, and was probably discarded when empty, few if any rule boxes are ever encountered.

Repairs

Generally speaking, broken or damaged rules have little if any value to collectors, either as examples in their collection, or as a source for parts for repairs.

It is almost never possible to repair a broken or damaged rule. Once a joint pin comes loose from the outer joint plates so it can be seen to turn when the joint is opened/closed, it is virtually impossible to tighten it or replace it, and any attempt to do so will leave the rule looking worse than before. If the edge binding comes loose it usually cannot be glued/re-riveted down and look right.

Nor can parts be switched between rules, even of the same type/maker. Replacing one stick or joint plate/assembly with another requires a repinning which leaves clear evidence of its presence. It is not even possible to switch caliper or through slides between rules. Slides were all individually fitted to the specific rule they ran in, and will not fit properly if switched to another rule . When slides and rules had to be graduated separately during manufacture, each was marked with a unique production number so it could be reunited with its proper counterpart after the separate operations (examination of any caliper rule will reveal these numbers stamped in the groove and on the bottom surface of the slide).

Cleaning and Refinishing

It is difficult, if not impossible, to clean a rule or refinish a rule without leaving evidence of having done so. Alcohol and lacquer thinner will dissolve the original finish on most rules, leaving them looking bleached and washed out. Many cleaning compounds contain oils such as lanolin or linseed oil, and these oils can get into the pores of the wood and leave it looking dark or stained. Mechanical cleaning is worse; steel wool leaves unmistakable evidence of its use, and even pencil erasers damage the wood/metal surfaces to some extent. The most that can be done is a gentle washing with slightly soapy water followed by thorough rinsing and immediate drying, and then a quick wipe down with petroleum distillate or turpentine to remove any oily surface dirt. Similarly, any attempt to refinish a rule with varnish, shellac, tung oil, etc. will leave it looking worse than before. None of these materials were used to finish the rule originally, and all present a different appearance from the original patinated lacquer.

WOOD FOLDING RULES

6 INCH, 2 FOLD WOOD RULES	A. Stanley 1854-1857	Stanley Rule &Level 1859-....	Stearns (early) 1853-1856	Stearns (late) 1856-1902	Chapin (early) 1835-185?	Chapin (late) 185?-1901
Round Joint, Narrow		30 $400-$800				
Round Joint			44 $125-$300	44 $125-$300		1½ $40-$150
Square Joint, Caliper Slide, Broad		13 $35-$125	22 $200-$450	13 $175-$400		70¾ $50-$175
		13½ $50-$150				70¼ $60-$200
	36 $600-$900	36 $10-$40		13½ $175-$400	36 $60-$250	70 $75-$200
			21 $220-$500	12 $200-$450		71 $85-$225
Square Joint, Full Bound, Caliper Slide, Broad		14 $350-$650				71½ $100-$250

1 FOOT, 2 FOLD WOOD RULES	A. Stanley 1854-1857	Stanley Rule&Level 1859-....	Stearns (early) 1853-1856	Stearns (late) 1856-1902	Chapin (early) 1835-185?	Chapin (late) 185?-1901
Square Joint, Caliper Slide		35 $250-$500				
Square Joint, Caliper Slide, Broad	36½ $650-$800	36½ $15-$45				70½ $50-$150
		36½L $15-$45				
		36½R $15-$45				
		36½B $450-$900				
Square Joint, Full Bound, Broad		036 $300-$600				

1 FOOT, 4 FOLD WOOD RULES	A. Stanley 1854-1857	Stanley Rule&Level 1859-....	Stearns (early) 1853-1856	Stearns (late) 1856-1902	Chapin (early) 1835-185?	Chapin (late) 185?-1901
Joint & Plates Unknown		188 $800-$1600				
Round Joint, Middle Plates, Narrow	69 $700-$850					
Round Joint, Middle Plates	69 $700-$850	69 $30-$100	43 $150-$350	43 $150-$350	69 $60-$200	1 $40-$150
			42 $175-$400	42 $175-$400		
Round Joint, Middle Plates, Broad				40 $160-$375		

Stephens 1854-1901	Chapin-Stephens 1901-1929	Belcher Bros. 1822-1877	Standard Rule 1872-1889	Upson Nut 1889-1922	Lufkin 1924-....	
		89 $40-$100				
	13 $40-$150					
100 $75-$200	13½ $40-$175			13½ $40-$140	172 $40-$140	Extra Wide
95 $75-$200	36 $50-$175	P99 $50-$130	36 $125-$300	36 $50-$140	171 $25-$125	
96 $100-$240	14 $75-$225	99 $75-$200				Cased Leg
	13¾ $40-$175					Through Slide
97 $125-$275	14½ $75-$225					

Stephens 1854-1901	Chapin-Stephens 1901-1929	Belcher Bros. 1822-1877	Standard Rule 1872-1889	Upson Nut 1889-1922	Lufkin 1924-....	
		89 $75-$200			371 $60-$200	
100 $40-$140	36½ $35-$125			36½ $30-$120	372 $30-$100	
						Right-to-Left Graduations, Right Hand Caliper
						Full Bound
36 $125-$275	036 $100-$250					Level and Blade in Legs

Stephens 1854-1901	Chapin-Stephens 1901-1929	Belcher Bros. 1822-1877	Standard Rule 1872-1889	Upson Nut 1889-1922	Lufkin 1924-....	
						Printers' Rule
		0 $75-$100				
70 $30-$125	69 $25-$110	10 $75-$100		69 $40-$140	465 $30-$125	
						'Selected'
		20 $50-$125				

1 FOOT, 4 FOLD WOOD RULES (continued)	A. Stanley 1854-1857	Stanley Rule&Level 1859-....	Stearns (early) 1853-1856	Stearns (late) 1856-1902	Chapin (early) 1835-185?	Chapin (late) 185?-1901
Square Joint, Middle Plates, Narrow	65 $800-$1000					
Square Joint, Middle Plates	65 $800-$1000	65 $75-$250		75 $200-$450	65 $60-$250	2 $50-$200
		165 $600-$1200				
Square Joint, Middle Plates, Broad						
Sqr. Joint, Middle Plates, Broad, Caliper Slide						
Square Joint, Edge Plates, Narrow	64 $800-$1000					
Square Joint, Edge Plates	64 $800-$1000	64 $75-$250			64 $125-$300	3 $100-$250
Square Joint, Edge Plates, Broad				39 $210-$475		
Square Joint, Half Bound, Narrow						
Square Joint, Half Bound						4 $125-$27?
Square Joint, Half Bound, Broad						
Square Joint, Full Bound, Narrow	65½ $900-$1100					
Square Joint, Full Bound	65½ $900-$1100	65½ $125-$300		74 $235-$550		5 $125-$30?
Square Joint, Full Bound, Caliper Slide		3 $150-$350				
Square Joint, Full Bound, Broad						
Square Joint, Full Bound, Broad, Caliper Slide						
Arch Joint, Middle Plates, Narrow	55 $900-$1100					
Arch Joint, Middle Plates	55 $900-$1100	55 $125-$300			55 $125-$300	6 $100-$25?
Arch Joint, Middle Plates, Broad						
Arch Joint, Middle Plates, Broad, Caliper Slide						
Arch Joint, Edge Plates, Narrow	56 $950-$1200					
Arch Joint, Edge Plates	56 $950-$1200	56 $125-$300			56 $140-$325	7 $100-$27?
Arch Joint, Edge Plates, Broad				30 $235-$550		
Arch Joint, Edge Plates, Broad, Caliper Slide		32 $15-$50		30½ $250-$600		72 $125-$32?

Stephens 1854-1901	Chapin-Stephens 1901-1929	Belcher Bros. 1822-1877	Standard Rule 1872-1889	Upson Nut 1889-1922	Lufkin 1924-....	
		1 $50-$120				
71 $75-$175	65 $50-$160	11 $50-$120		65 $40-$150	475 $25-$125	Metric & English Graduations
		21 $60-$140				
		C21 $75-$175				
		2 $50-$120				
72 $75-$225	64 $60-$210	12 $50-$120		64 $50-$175		
72¼ $90-$275	64¼ $90-$250	22 $60-$140				
		3 $60-$150				
		13 $60-$150	64½ $160-$350	64½ $70-$190		
		23 $75-$175				
		4 $75-$175				
72½ $100-$275	65½ $90-$250	14 $75-$175		65½ $60-$210	478 $75-$180	
	3 $100-$275					
75½ $100-$275		24 $75-$200				
		C24 $100-$250				
		5 $60-$140				
73 $90-$240	55 $80-$225	15 $60-$140		55 $50-$175		
		25 $75-$175				
		C25 $90-$225				
		6 $60-$140				
74 $80-$250	56 $75-$225	16 $60-$140		56 $50-$200		
		26 $75-$175		56¼ $70-$210		
98 $125-$300	32 $75-$275			32 $75-$225	386 $70-$210	

13

1 FOOT, 4 FOLD WOOD RULES (continued)	A. Stanley 1854-1857	Stanley Rule&Level 1859-....	Stearns (early) 1853-1856	Stearns (late) 1856-1902	Chapin (early) 1835-185?	Chapin (late) 185?-1901
Arch Joint, Half Bound, Narrow						
Arch Joint, Half Bound						8 $150-$350
Arch Joint, Half Bound, Broad						
Arch Joint, Full Bound, Narrow	57 $900-$1200					
Arch Joint, Full Bound	57 $900-$1200	57 $175-$350 / 57 $750-$1100			57 $150-$350	9 $125-$300
Arch Joint, Full Bound, Broad				29 $240-$575		
Arch Joint, Full Bound, Broad, Caliper Slide		32½ $25-$75		29½ $260-$625		73 $150-$350
Double Arch Joint, Middle Plates, Narrow						
Double Arch Joint, Middle Plates						
Double Arch Joint, Middle Plates, Broad						
Double Arch Joint, Edge Plates, Narrow						
Double Arch Joint, Edge Plates						
Double Arch Joint, Edge Plates, Broad						
Double Arch Joint, Full Bound, Narrow						
Double Arch Joint, Full Bound						
Double Arch Joint, Full Bound, Broad						

2 FOOT, 2 FOLD WOOD RULES	A. Stanley 1854-1857	Stanley Rule&Level 1859-....	Stearns (early) 1853-1856	Stearns (late) 1856-1902	Chapin (early) 1835-185?	Chapin (late) 185?-1901
Round Joint	29 $800-$900	29 $75-$200			29 $75-$225	38 $50-$175
Square Joint	18 $400-$900	18 $20-$50	11 $230-$450	11 $230-$450	18 $125-$300	39 $100-$250
	19 $1300-$1800	19 $500-$900	10 $350-$750	10 $350-$750	19 $125-$300	
	22 $1100-$1500	22 $250-$500			22 $180-$375	40 $140-$325
				111 $900-$1500		

Stephens 1854-1901	Chapin-Stephens 1901-1929	Belcher Bros. 1822-1877	Standard Rule 1872-1889	Upson Nut 1889-1922	Lufkin 1924-....	
		7 $75-$175				
		17 $75-$175		56½ $90-$225		
		27 $100-$225				
		8 $75-$200				
75 $100-$275	57 $100-$250	18 $75-$200	57 $175-$400	57 $60-$210		German Silver Joints & Trim
	57½ $85-$275	28 $100-$250				
99 $125-$325	32½ $125-$300	C28 $125-$300		32½ $75-$225	388 $60-$200	
		9 $125-$300				
		19 $125-$300				
		29 $150-$350				
		E9 $125-$300				
		E19 $125-$300				
		E29 $150-$350				
		B9 $175-$375				
		B19 $175-$375				
		B29 $200-$425				

Stephens 1854-1901	Chapin-Stephens 1901-1929	Belcher Bros. 1822-1877	Standard Rule 1872-1889	Upson Nut 1889-1922	Lufkin 1924-....	
1 $50-$160	29 $40-$150	80 $75-$175		29 $50-$125	602 $40-$110	
2 $75-$225	18 $60-$200	81 $75-$190		18 $75-$175	703 $40-$160	
		82 $75-$190				
5 $125-$300	22 $85-$275		22 $200-$425	22 $85-$225		Board Tables
						Spring Joint

2 FOOT, 2 FOLD WOOD RULES (continued)	A. Stanley 1854-1857	Stanley Rule&Level 1859-....	Stearns (early) 1853-1856	Stearns (late) 1856-1902	Chapin (early) 1835-185?	Chapin (late) 185?-1901
Square Joint (continued)					20 $110-$275	
					21 $175-$400	
					23 $225-$450	
Square Joint, Plain Slide	26 $800-$1100	26 $100-$300		9 $220-$475		46 $60-$200
Square Joint, Gunter's Slide	27 $850-$1200	27 $125-$350	5(*) $325-$600	8 $300-$550	26 $125-$300	47 $100-$250
			6(*) $300-$550		25 $125-$300	
					27 $175-$400	
Square Joint, Full Bound	28 $1300-$1700	28 $500-$800				
					24 $175-$350	
Square Joint, Full Bound, Gunter's Slide					28 $225-$450	
Arch Joint	1 $850-$1100	1 $100-$300	7(*) $275-$525	6 $220-$475	1 $100-$250	41 $50-$200
	2 $900-$1200	2 $150-$400	8(*) $300-$550	7 $225-$500	2 $110-$275	42 $75-$225
		4 $300-$600		6½ $475-$800	3 $240-$450	44 $160-$400
	7 $1600-$2000	7 $700-$1200			7 $240-$500	43 $175-$450
		101 $800-$1300				
				107 $1200-$1800		
					8 $350-$700	
Arch Joint, Plain Slide						

Stephens 1854-1901	Chapin-Stephens 1901-1929	Belcher Bros. 1822-1877	Standard Rule 1872-1889	Upson Nut 1889-1922	Lufkin 1924-....	
		83 $200-$450				Extra Thin
						Board Tables
						Extra Thin, Board Tables
		80 $75-$185				No Maker's Name
					702 $50-$160	
9 $60-$200	26 $75-$175	P91 $125-$275	26 $100-$275	26 $40-$140		
	27 $70-$210	92 $125-$300	27 $140-$325	27 $50-$175		Carpenters' Sliding Rule
		91 $125-$300				Carpenters' Sliding Rule
						Carpenters' Sliding Rule, Extra Thin
		90 $100-$250				No Maker's Name
		93 $200-$450				Engineers' Sliding Rule
		84 $125-$300				
						Board Tables
		94 $150-$350				Carpenters' Sliding Rule
	1 $75-$175	85 $100-$250	1 $110-$275	1 $40-$150	802 $25-$125	
13 $75-$210	2 $50-$190	86 $100-$250	2 $125-$300	2 $45-$160		
15½ $175-$375	3½ $150-$350			4 $125-$275		Extra Thin
18 $150-$425	23 $160-$380	586 $150-$350				Board Tables
						Metric and English Graduations
15 $60-$200		87 $100-$250				
						Spring Joint
						Extra Thin, Board Tables
23 $75-$225	11 $60-$200					

2 FOOT, 2 FOLD WOOD RULES (continued)	A. Stanley 1854-1857	Stanley Rule&Level 1859-....	Stearns (early) 1853-1856	Stearns (late) 1856-1902	Chapin (early) 1835-185?	Chapin (late) 185?-1901
Arch Joint, Gunter's Slide	6 $1300-$1800	6 $450-$900	1 $750-$1200	2 $650-$1100	16 $425-$850	50 $300-$80
		6 $800-$1400				
	12 $900-$1200	12 $150-$400	3(*) $300-$650	4 $275-$600	11 $175-$400	48 $150-$35
			4(*) $260-$550	5 $240-$500		
					12 $175-$400	
					13 $275-$600	
Arch Joint, Arch Tips					4 $300-$550	
					9 $350-$700	
Arch Joint, Arch Tips, Gunter's Slide					14 $550-$1000	
Arch Joint, Half Bound	3 $1300-$1800	3 $450-$900				
Arch Joint, Half Bound, Gunter's Slide	14 $1300-$1800	14 $450-$900				
Arch Joint, Full Bound	5 $800-$1100	5 $75-$300			5 $225-$450	45 $160-$40
					6 $275-$600	
Arch Joint, Full Bound, Arch Tips					10 $450-$850	
Arch Joint, Full Bound, Gunter's Slide	15 $850-$1200	15 $150-$400		3 $275-$600	15 $200-$475	49 $175-$42
	16 $1300-$1800	16 $450-$900		1 $650-$1100	17 $450-$850	51 $300-$80
		16 $800-$1400				
Fancy Joint						44½ $200-$50

2 FOOT, 4 FOLD WOOD RULES	A. Stanley 1854-1857	Stanley Rule&Level 1859-....	Stearns (early) 1853-1856	Stearns (late) 1856-1902	Chapin (early) 1835-185?	Chapin (late) 185?-190
Round Joint, Plates Unknown				33 $140-$325		
Round Joint, Middle Plates, Narrow				44 $140-$325		

Stephens 1854-1901	Chapin-Stephens 1901-1929	Belcher Bros. 1822-1877	Standard Rule 1872-1889	Upson Nut 1889-1922	Lufkin 1924-....	
16 $325-$760	**6** $300-$675	**97** $250-$550	**6** $400-$900	**6** $200-$550		Engineers' Sliding Rule
						Engineers' Sliding Rule w/Jillson's H.P. Table
14 $140-$325	**12** $125-$300	**96** $175-$400	**12** $175-$450	**12** $100-$250		Carpenters' Sliding Rule
						Carpenters' Sliding Rule
						Carpenters' Sliding Rule
						Carpenters' Sliding Rule, Extra Thin
		95 $175-$400				
						Board Tables
						Carpenters' Sliding Rule
						Carpenters' Sliding Rule
17 $160-$375	**5** $150-$350	**88** $125-$300	**5** $190-$500	**5** $100-$275	**308** $90-$250	
22 $200-$500	**24** $175-$450					Board Tables
						Board Tables
27 $175-$400	**15** $150-$350	**98** $200-$450	**15** $200-$575	**15** $125-$300		Carpenters' Sliding Rule
28 $325-$750	**16** $300-$675		**16** $400-$900	**16** $200-$550		Engineers' Sliding Rule
						Engineers' Sliding Rule w/Jillson's H.P. Table
15½ $180-$460	**4** $175-$425					Extra Thin

Stephens 1854-1901	Chapin-Stephens 1901-1929	Belcher Bros. 1822-1877	Standard Rule 1872-1889	Upson Nut 1889-1922	Lufkin 1924-....	
		40 $40-$100				
		30 $50-$130				Extra Thick

2 FOOT, 4 FOLD WOOD RULES (continued)	A. Stanley 1854-1857	Stanley Rule&Level 1859-....	Stearns (early) 1853-1856	Stearns (late) 1856-1902	Chapin (early) 1835-185?	Chapin (late) 185?-1901
Round Joint, Middle Plates	**68** $600-$800	**68** $10-$35	**41** $150-$325	**41** $140-$325	**68** $50-$175	**10** $25-$125
		8 $450-$900				
		68A $10-$35				
		(No #) $75-$150				
		27 $10-$25				
		163 $10-$30				
Round Joint, Middle Plates, Broad	**67** $800-$1000	**67** $75-$275			**67** $75-$225	**22** $50-$175
Square Joint, Plates Unknown				**33½** $150-$350		
Square Joint, Plates Unknown, Broad				**31** $150-$350		
				32½ $140-$340		
Square Joint, Middle Plates, Narrow	**61½** $700-$1000	**61½** $40-$125		**36** $140-$340		**11½** $40-$150
Square Joint, Middle Plates	**61** $650-$850	**61** $15-$35	**38** $175-$400	**38** $150-$350	**61** $60-$200	**11** $40-$150
		161 $600-$1200				
		61A $20-$60				
			39 (*) $175-$400			
			40 (*) $175-$400			
				34½ $150-$350		
					62 $60-$200	
Square Joint, Middle Plates, Broad	**70** $650-$950	**70** $35-$150		**32** $200-$400	**70** $75-$225	**23** $50-$175
	71 $800-$1100				**71** $75-$225	
		7 $40-$200				

Stephens 1854-1901	Chapin-Stephens 1901-1929	Belcher Bros. 1822-1877	Standard Rule 1872-1889	Upson Nut 1889-1922	Lufkin 1924-....	
41 $25-$125	68 $25-$100	60 $40-$100	68 $35-$160	68 $25-$90	651 $20-$80	
		50 $50-$130	8 $60-$200			Extra Thick
	68P $30-$125					Printed Graduations & Figures
						"4-Square" Rule
					48 $15-$75	Maple
						Steel Joints & Trim
53 $45-$160	67 $40-$150	70 $60-$150	67 $80-$225	67 $30-$125		
						Board Tables
44 $35-$140	61½ $25-$125	41 $60-$150	61½ $60-$200	61½ $25-$100		
		31 $75-$175				Extra Thick
42 $35-$140	61 $25-$125	61 $60-$150	61 $60-$200	61 $25-$110	751 $20-$100	
						Metric & English Graduations
	61P $35-$100				751B $25-$100	Printed Graduations & Figures
		51 $75-$175				Extra Thick
54 $45-$160	70 $45-$150	71 $90-$225	70 $100-$240	70 $25-$125	752 $25-$110	
					752B $75-$175	Blindman's Markings

2 FOOT, 4 FOLD WOOD RULES (continued)	A. Stanley 1854-1857	Stanley Rule&Level 1859-....	Stearns (early) 1853-1856	Stearns (late) 1856-1902	Chapin (early) 1835-185?	Chapin (late) 185?-1901
Square Joint, Middle Plates (continued)		70E $75-$250				
Square Joint, Edge Plates, Narrow		63½ $50-$150		33 $160-$375		13 $60-$200
Square Joint, Edge Plates	63 $700-$900	63 $25-$100	37 $175-$400	37 $175-$400 137 $900-$1500	63 $100-$250	12 $60-$200
Square Joint, Edge Plates, Broad	72 $800-$1200 79 $950-$1300	72 $75-$300 79 $225-$450 7 $40-$200	29 (*) $200-$450 30 (*) $200-$450		72 $120-$275 79 $160-$375	24 $75-$225 34 $125-$325 34½ $175-$425
Square Joint, Edge Plates, Broad, T Hook						34¾ $175-$425
Square Joint, Half Bound, Narrow						
Square Joint, Half Bound		84 $20-$50				14 $75-$175
Square Joint, Half Bound, Broad		72¼ 500-$1000				25 $100-$250
Square Joint, Full Bound, Narrow		62½ $40-$125				15½ $60-$175
Square Joint, Full Bound	62 $650-$850	62 $35-$70		35 $160-$375 135 $1000-$1600		15 $75-$175

Stephens 1854-1901	Chapin-Stephens 1901-1929	Belcher Bros. 1822-1877	Standard Rule 1872-1889	Upson Nut 1889-1922	Lufkin 1924-....	
						Printed Graduations & Figures
		571 $150-$350				Board Scales
44½ $50-$180	**63½** $50-$160	**42** $60-$150	**63½** $100-$275	**63½** $30-$140	**760** $25-$130	
		32 $75-$175				Extra Thick
45 $50-$180	**63** $50-$175	**62** $60-$150	**63** $100-$275	**63** $30-$140	**761** $25-$130	Spring Joint
		52 $75-$175				Extra Thick
56 $60-$200	**72** $60-$200	**72** $100-$225	**72** $125-$300	**72** $45-$160	**762** $40-$150	
64 $125-$300	**79** $125-$275		**79** $200-$450	**79** $75-$225		Board Tables
					762B $50-$175	Blindman's Markings
	79½ $160-$375					Board Scales
	7 $100-$225					Red Blindman's Markings
(No #) $175-$400	**79¾** $150-$375					Board Scales
		43 $75-$180				
		33 $75-$200				Extra Thick
42¼ $50-$160	**84** $40-$150	**63** $75-$180	**84** $100-$240	**84** $25-$125	**771** $20-$115	
		53 $75-$200				Extra Thick
54½ $100-$240	**72¼** $75-$220	**73** $90-$225	**72¼** $140-$340	**72¼** $50-$175		
42¾ $50-$160	**62½** $40-$150	**44** $75-$200	**62½** $75-$225	**62½** $25-$125	**780** $20-$110	
		34 $100-$250				Extra Thick
42½ $50-$160	**62** $40-$150	**64** $75-$200	**62** $75-$225	**62** $25-$125	**781** $20-$110	Spring Joint
						Spring Joint
		54 $100-$250				Extra Thick

2 FOOT, 4 FOLD WOOD RULES (continued)	A. Stanley 1854-1857	Stanley Rule&Level 1859-....	Stearns (early) 1853-1856	Stearns (late) 1856-1902	Chapin (early) 1835-185?	Chapin (late) 185?-1901
Square Joint, Full Bound, Caliper Slide		62C $200-$400				
Square Joint, Full Bound, Broad	72½ $750-$1000	72½ $75-$200				26 $100-$250
	80 $1300-$1800	80 $500-$1000			80 $250-$500	35 $250-$500
Square Joint, Full Bound, Broad, Caliper Slide		72½ $500-$1000				
Arch Joint, Middle Plates, Narrow						
Arch Joint, Middle Plates	51 $650-$850	51 $25-$75		46 $160-$375	51 $75-$225	16 $50-$175
		151 $700-$1200				
				27 $160-$375	52 $75-$225	
Arch Joint, Middle Plates, Broad	73 $800-$1100	73 $75-$300		21 $225-$450	73 $75-$225	27 $50-$175
		73¼ $150-$350				
	74 $500-$1000				74 $75-$225	
		173 $800-$1200				
Arch Joint, Edge Plates, Narrow				31 $200-$400		
						13½ $75-$22
Arch Joint, Edge Plates	53 $600-$1000	53 $45-$100	34 (*) $260-$500	45 $225-$450	53 $110-$275	17 $75-$22
		53½ $40-$125				17½ $125-$27
			35 (*) $260-$500			
				26 $400-$750		
Arch Joint, Edge Plates, Broad	75 $700-$1100	75 $100-$300	25 (*) $280-$600	20 $300-$550	75 $150-$350	28 $140-$30
	81 $1400.$2000	81 $275-$550		18 $550-$900	81 $250-$500	36 $225-$45

Stephens 1854-1901	Chapin-Stephens 1901-1929	Belcher Bros. 1822-1877	Standard Rule 1872-1889	Upson Nut 1889-1922	Lufkin 1924-....	
	62C $125-$300				**781C** $100-$250	
54½ $75-$225	**72½** $75-$220	**74** $100-$250	**72½** $140-$325	**72½** $50-$175	**782** $45-$160	
65 $150-$340	**79¼** $125-$300					Board Tables
		45 $75-$200				
		35 $100-$240				Extra Thick
46 $45-$160	**51** $40-$150	**65** $75-$200	**51** $100-$240	**51** $25-$125	**851** $25-$110	
						Metric & English Graduations
		55 $100-$240				Extra Thick
57 $45-$160	**73** $40-$150	**75** $125-$275	**73** $100-$240	**73** $25-$125	**852** $25-$120	
						Metric & English Graduations
44¾ $75-$200	**53¼** $50-$175	**46** $75-$200				
		36 $100-$240				Extra Thick
	53¾ $60-$180					Inside Beveled Edges
48 $75-$220	**53** $60-$180	**66** $75-$200	**53** $125-$300	**53** $50-$160	**861** $40-$150	Inside Beveled Edges
39 $100-$260	**53½** $75-$225		**53½** $160-$375	**53½** $60-$200	**861A** $50-$175	
						Board Tables
		56 $100-$240				Extra Thick
59 $125-$275	**75** $100-$250	**76** $125-$275	**75** $175-$400	**75** $60-$200	**862** $60-$200	
66 $160-$375	**81** $150-$350		**81** $225-$540	**81** $120-$280		Board Tables

2 FOOT, 4 FOLD WOOD RULES (continued)	A. Stanley 1854-1857	Stanley Rule&Level 1859-....	Stearns (early) 1853-1856	Stearns (late) 1856-1902	Chapin (early) 1835-185?	Chapin (late) 185?-1901
Arch Joint, Edge Plates, Broad (continued)			26 (*) $300-$550 27 (*) $300-$550	118 $1200-$1800		28½ $140-$325
Arch Joint, Edge Plates, Broad, Plain Slide		83 $350-$700				33 $200-$450
Arch Joint, Edge Plates, Broad, Gunter's Slide		83 $600-$1200	28 $1100-$1500	19 $1000-$1400		
Arch Joint, Edge Plates, Broad, Caliper Slide		83¼ $500-$1000 83½ $500-$1000 83C $300-$600				33½ $180-$425
Arch Joint, Half Bound, Narrow						
Arch Joint, Half Bound		52 $100-$250				18 $140-$325
Arch Joint, Half Bound, Broad						29 $150-$350
Arch Joint, Full Bound, Narrow				72 $275-$500		
Arch Joint, Full Bound	54 $550-$950	54 $40-$80	32 $325-$600 31 $425-$800	23 $300-$550 22 $400-$750	54 $225-$400	19 $150-$350

Stephens 1854-1901	Chapin-Stephens 1901-1929	Belcher Bros. 1822-1877	Standard Rule 1872-1889	Upson Nut 1889-1922	Lufkin 1924-....	
						Spring Joint
	75½ $125-$300					Inside Beveled Edges
		576 $175-$450				Board Scales
					863L $100-$250	Protractor Joint, Level in Edge
	83 $160-$375	9076 $250-$600		83 $140-$320		
			83 $300-$700	83 $210-$475		
	83½ $150-$360				862C $110-$280	
		47 $75-$200				
		37 $100-$240				Extra Thick
49½ $125-$300	52 $110-$275	67 $75-$200	52 $200-$450	52 $75-$225		
		57 $100-$240				Extra Thick
??? $125-$300						Inside Beveled Edges
60¼ $140-325	76½ $125-$300	77 $110-$275				
					873L $100-$250	Protractor Joint, Level In Edge
		48 $100-$250				
		38 $125-$300				Extra Thick
49 $140-$325	54 $125-$300	68 $100-$250	54 $200-$475	54 $100-$250	881 $75-$225	Board Tables
??? $150-$350	54S $125-$300				881D $100-$240	100ths of a Foot Graduations
		58 $125-$300				Extra Thick

2 FOOT, 4 FOLD WOOD RULES (continued)	A. Stanley 1854-1857	Stanley Rule&Level 1859-....	Steams (early) 1853-1856	Stearns (late) 1856-1902	Chapin (early) 1835-185?	Chapin (late) 185?-1901
Arch Joint, Full Bound, Broad	76 $650-$1100	76 $80-$160		15 $325-$600	76 $180-$400	30 $160-$375
	82 $1000.$1600	82 $300-$600	23 $500-$850	14 $425-$800	82 $300-$550	37 $275-$500
Arch Joint, Full Bound, Broad, Caliper Slide		76C $200-$400				
		76½ $500-$900				
Double Square Joint, Edge Plates			36 $2000-$3000	34 $2000.$3000		
Double Arch Joint, Middle Plates, Narrow						
Double Arch Joint, Middle Plates	59 $1100.$1700	59 $180-$350			59 $160-$425	20 $160-$375
					58 $160-$425	
Double Arch Joint, Middle Plates, Broad	77 $800-$1400	77 $200-$400			77 $225-$450	31 $175-$400
					83 $375-$700	
Double Arch Joint, Edge Plates, Narrow						
Double Arch Joint, Edge Plates	59 $800-$1200	59 $180-$350	33 $375-$700	25 $350-$650		
				24 $600-$900		
				125 $1400-$2200		
Double Arch Joint, Edge Plates, Broad	77 $900-$1300	77 $200-$400		17 $400-$750		
			24 $625-$1000	16 $600-$950		
Double Arch Joint, Half Bound, Broad	78 $900-$1300	78 $200-$400				32½ $200-$450
Double Arch Joint, Full Bound, Narrow						
Double Arch Joint, Full Bound	60 $900-$1300	60 $200-$400			60 $240-$475	21 $180-$425

Stephens 1854-1901	Chapin-Stephens 1901-1929	Belcher Bros. 1822-1877	Standard Rule 1872-1889	Upson Nut 1889-1922	Lufkin 1924-....	
60 $160-$350	**76** $140-$325	**78** $150-$350	**76** $220-$500	**76** $100-$250	**882** $90-$240	
67 $180-$425	**82** $160-$375		**82** $250-$600	**82** $125-$300		Board Tables
	76C $180-$425					
		49 $150-$350				
		39 $175-$400				Extra Thick
50 $150-$350	**59** $140-$325	**69** $150-$350	**59** $220-$500	**59** $90-$250	**951** $90-$240	
		59 $175-$400				Extra Thick
61 $160-$380	**77** $150-$340	**79** $200-$450	**77** $240-$540	**77** $110-$275		
						Board Tables
		E49 $150-$350				
		E39 $175-$400				Extra Thick
		E69 $150-$350				
						Board Tables
						Spring Joint
						Extra Thick
		E59 $175-$400				
		E79 $200-$450				
						Board Tables
	78 $160-$375		**78** $250-$600	**78** $125-$300		
		B49 $180-$425				
		B39 $225-$500				Extra Thick
52 $175-$400	**60** $150-$360	**B69** $180-$425	**60** $240-$575	**60** $125-$300		

29

2 FOOT, 4 FOLD WOOD RULES (continued)	A. Stanley 1854-1857	Stanley Rule&Level 1859-....	Stearns (early) 1853-1856	Stearns (late) 1856-1902	Chapin (early) 1835-185?	Chapin (late) 185?-1901
Double Arch Joint, Full Bound,	78½ $800-$1200	78½ $180-$350			78 $300-$550	32 $220-$500
					84 $425-$800	

3 FOOT, 4 FOLD WOOD RULES	A. Stanley 1854-1857	Stanley Rule&Level 1859-....	Stearns (early) 1853-1856	Stearns (late) 1856-1902	Chapin (early) 1835-185?	Chapin (late) 185?-1901
Round Joint, Middle Plates						83¼ $50-$175
Round Joint, Middle Plates, Broad						83½ $60-$200
Square Joint, Middle Plates	66 $900-$1300				66 $100-$250	
		66½ $10-$30				23¼ $60-$200
		133 $100-$300				
Square Joint, Middle Plates, Broad		170BE $180-$350				
						23½ $60-$200
Square Joint, Edge Plates		8 $300-$600				
Arch Joint, Middle Plates	66 $900-$1300	66 $200-$400		28 $300-$600		84 $125-$300
		66½ $50-$100				84¼ $75-$225
		66½A $25-$75				
Arch Joint, Middle Plates, Broad		173E $200-$500				
		173¼E $200-$500				
						85¼ $60-$200
Arch Joint, Edge Plates		66¼ $30-$80				
		66¼A $100-$300				
Arch Joint, Full Bound		66¾ $50-$150				84¾ $75-$225
		66B $250-$500				
						84½ $150-$350
Double Arch Joint, Plates Unknown, Broad						85¾ $200-$450

Stephens 1854-1901	Chapin-Stephens 1901-1929	Belcher Bros. 1822-1877	Standard Rule 1872-1889	Upson Nut 1889-1922	Lufkin 1924-....	
63 $200-$475	**78½** $180-$425	**B79** $225-$500	**78½** $275-$675	**78½** $150-$350	**982** $125-$300	
66 $250-$600						Board Tables

Stephens 1854-1901	Chapin-Stephens 1901-1929	Belcher Bros. 1822-1877	Standard Rule 1872-1889	Upson Nut 1889-1922	Lufkin 1924-....	
	68¼ $40-$150					
	67¼ $50-$175					
	32 $100-$250					Fractions of a Yard
	61¼ $50-$175				**3851** $25-$125	
						Maple, Printed Graduations and Figures
					3752B $30-$140	Blindman's Markings
	70¼ $45-$175					
					3761B $25-$125	Blindman's Markings
	66 $100-$250	**351** $125-$300	**66** $175-$400	**66** $60-$200		Fractions of a Yard
	66½ $60-$200			**66½** $45-$160	**3851** $25-$125	
32 $40-$150						Printed Graduations and Figures
	73¼ $175-$175					
					3861 $25-$125	
	66¼A $50-$175				**66¼A** $25-$125	Printed Graduations and Figures
	66¾ $60-$200				**3881** $40-$150	
	66¼ $125-$300					Fractions of a Yard
	77¼ $160-$375					

4 FOOT, 4 FOLD WOOD RULES	A. Stanley 1854-1857	Stanley Rule&Level 1859-....	Stearns (early) 1853-1856	Stearns (late) 1856-1902	Chapin (early) 1835-185?	Chapin (late) 185?-1901
Arch Joint, Edge Plates						
Arch Joint, Full Bound		94 $75-$300				96½ $220-$500

MISCELLANEOUS WOOD FOLDING RULES	A. Stanley 1854-1857	Stanley Rule&Level 1859-....	Stearns (early) 1853-1856	Stearns (late) 1856-1902	Chapin (early) 1835-185?	Chapin (late) 185?-1901
20 Cm., 2 Fold, Square Joint, Caliper Slide		??? $400-$750				
8 Inch, 2 Fold, Square Joint, Caliper Slide		??? $500-$1000				
1 Foot, 3 Fold, Edge Plates, Narrow						95 $220-$500
1 Foot, 3 Fold, Half Bound, Narrow						
40 Cm., 4 Fold, Square Joint, Middle Plates		??? $300-$650				
40 Cm., 4 Fold, Arch Joint, Full Bound		??? $600-$1200				
1/2 Meter, 4 Fold, Arch Joint, Middle Plates		10 $400-$800				
2 Foot, 2 Fold, Edge Plates		(No #) $2000-$4000		(No #) $2000-$4000		
2 Foot, 3 Fold, Edge Plates						
2 Foot, 3 Fold, Half Bound						
2 Foot, 6 Fold, Arch Joint, Middle Plates, Narrow						
2 Foot, 6 Fold, Arch Joint, Edge Plates, Narrow		58 $300-$600 / 58½ $3000-$6000		28½ $500-$900		96 $220-$500
2 Foot, 6 Fold, Arch Joint, Full Bound, Narrow		58½ $400-$750				96¼ $250-$600
Extension Stick, 2 to 4 Foot, 3 Fold						
1 Meter, 4 Fold, Arch Joint, Middle Plates		20 $400-$800				
1 Meter, 4 Fold, Arch Joint, Middle Plates, Broad		30 $350-$700				
1 Meter, 5 Fold, Middle Plates		??? $3000-$6000				

Stephens 1854-1901	Chapin-Stephens 1901-1929	Belcher Bros. 1822-1877	Standard Rule 1872-1889	Upson Nut 1889-1922	Lufkin 1924-....	
??? $100-$250		476 $125-$300				
	94 $90-$250				4883 $60-$200	"Carriage Maker's"

Stephens 1854-1901	Chapin-Stephens 1901-1929	Belcher Bros. 1822-1877	Standard Rule 1872-1889	Upson Nut 1889-1922	Lufkin 1924-....	
						Metric Graduations
	58¾ $180-$425					
					2061 $125-$300	Level in Edge
					2071 $150-$350	Level in Edge
						Metric Graduations
						Metric Graduations
						Metric and English Graduations
						Gear Calculating Rule
					2062 $175-$350	Level in Edge
					2072 $200-$450	Level in Edge
					2072P $450-$1000	Level in Edge, Plumb in Face
		S45 $375-$800				
33 $230-$525	58 $200-$450	S46 $375-$800	58 $325-$750	58 $160-$375		
						Weights of Metals Table
	58½ $220-$500		58½ $350-$800	58½ $180-$425		
	98 $150-$350					
						Metric and English Graduations
						Metric and English Graduations
						Metric and English Graduations

IVORY FOLDING RULES

6 INCH, 2 FOLD IVORY RULES	A. Stanley 1854-1857	Stanley Rule&Level 1859-....	Stearns (early) 1853-1856	Stearns (late) 1856-1902	Chapin (early) 1835-185?	Chapin (late) 185?-1901
Round Joint, Narrow	93 $1200-$1600	93 $350-$700	56(*) $400-$750			
		93½ $400-$800				
	93 $1200-$1600	93 $350-$700	57(*) $400-$750	61 $375-$700	92 $225-$350	68 $125-$300
Square Joint, Caliper Slide, Broad	37 $1400-$1800	37 $500-$1000		56 $350-$650	37 $225-$450	
	38 $900-$1300	38 $200-$400		55 $350-$650		74 $175-$400
			52(*) $400-$750			
				55½ $400-$750		75 $200-$450
Square Joint, Full Bound, Caliper Slide		40½ $250-$450				77½ $150-$350
Square Joint, Full Bound, Caliper Slide, Broad			52B $425-$800	55B $400-$750		74½ $175-$400
				56B $400-$750		

1 FOOT, 2 FOLD IVORY RULES	A. Stanley 1854-1857	Stanley Rule&Level 1859-....	Stearns (early) 1853-1856	Stearns (late) 1856-1902	Chapin (early) 1835-185?	Chapin (late) 185?-1901
Square Joint, Full Bound, Broad						
Arch Joint, Narrow						69 $250-$600

1 FOOT, 4 FOLD IVORY RULES	A. Stanley 1854-1857	Stanley Rule&Level 1859-....	Stearns (early) 1853-1856	Stearns (late) 1856-1902	Chapin (early) 1835-185?	Chapin (late) 185?-1901
Round Joint, Middle Plates, Narrow		90 $250-$500				
	90 $1000-$1500	90 $250-$500		59 $300-$600		52 $150-$350
				60 $300-$600		
Round Joint, Middle Plates	90 $1000-$1500	90 $250-$500				
	90 $1000-$1500	90 $250-$500			90 $200-$400	
Round Joint, Middle Plates, Broad						

Stephens 1854-1901	Chapin-Stephens 1901-1929	Belcher Bros. 1822-1877	Standard Rule 1872-1889	Upson Nut 1889-1922	Lufkin 1924-....	
34 $110-$275		289 $150-$325				
33 $110-$275	98½ $100-$250	189 $140-$300	93 $175-$400	93 $100-$250		Brass Joints & Trim
		P199 $150-$325				Brass Joints & Trim
95½ $160-$375	38 $140-$340	P299 $160-$350	38 $240-$540	38 $140-$325		
		199 $190-$425				Brass Joints & Trim, Cased Leg
96½ $190-$425	40¾ $160-$375	299 $200-$450				Cased Leg
	40½ $125-$300		40½ $200-$475	40½ $120-$275		
97½ $160-$375	40¼ $150-$350					
						Brass Joints & Trim

Stephens 1854-1901	Chapin-Stephens 1901-1929	Belcher Bros. 1822-1877	Standard Rule 1872-1889	Upson Nut 1889-1922	Lufkin 1924-....	
38 $2600-$5000						Blade & Level in Legs
99½ $240-$575	99½ $220-$500					

Stephens 1854-1901	Chapin-Stephens 1901-1929	Belcher Bros. 1822-1877	Standard Rule 1872-1889	Upson Nut 1889-1922	Lufkin 1924-....	
89½ $140-$325		200 $190-$425				
89 $140-$300	90 $125-$300	100 $175-$400	90 $220-$475	90 $120-$275		Brass Joints & Trim
						Brass Joints & Trim
		210 $190-$425				
		110 $175-$400		90 $120-$275		Brass Joints & Trim
		220 $200-$450				
		120 $190-$425				Brass Joints & Trim

1 FOOT, 4 FOLD IVORY RULES (continued)	A. Stanley 1854-1857	Stanley Rule&Level 1859-....	Steams (early) 1853-1856	Steams (late) 1856-1902	Chapin (early) 1835-185?	Chapin (late) 185?-1901
Round Joint, Edge Plates	90½ $1100-$1600					
Square Joint, Middle Plates, Narrow	92½ $1000-$1400	92½ $250-$450				54 $140-$32▋
	92½ $1000-$1400	92½ $250-$450				53 $140-$32▋
Square Joint, Middle Plates						
Square Joint, Middle Plates, Broad						
Square Joint, Edge Plates, Narrow	92 $900-$1300					
	92 $900-$1300					
Square Joint, Edge Plates		92 $200-$400	54(*) $350-$650	57 $325-$600		55 $150-$35▋
		92 $200-$400	55(*) $350-$650	58 $325-$600		
Square Joint, Edge Plates, Broad	91 $1100-$1500	91 $350-$600	53(*) $375-$700	52 $350-$650	91 $175-$400	
	91 $1100-$1500	91 $350-$600				
Square Joint, Edge Plates, Broad, Caliper Slide	39 $900-$1300	39 $200-$400	51 $400-$750	54 $375-$700		76 $160-$37▋
		39B $400-$800				
Square Joint, Full Bound, Narrow						58 $150-$35▋
Square Joint, Full Bound				57B $375-$700		58 $150-$35▋
Square Joint, Full Bound, Caliper Slide	40 $700-$1100	40 $175-$250				77 $200-$45▋
Square Joint, Full Bound, Broad			53B(*) $425-$800	52B $400-$750		58 $175-$40▋

Stephens 1854-1901	Chapin-Stephens 1901-1929	Belcher Bros. 1822-1877	Standard Rule 1872-1889	Upson Nut 1889-1922	Lufkin 1924-....
90½ $125-$300	92½ $120-$275	201 $200-$450			
90 $125-$300	90½ $120-$275	101 $180-$425			Brass Joints & Trim
		211 $200-$450	92½ $180-$425	92½ $120-$275	
		111 $180-$425			Brass Joints & trim
		221 $200-$480			
		121 $200-$450			Brass Joints & Trim
		202 $200-$450			
		102 $180-$425			Brass Joints & Trim
91 $140-$325	92 $125-$300	212 $200-$450	92 $200-$475	92 $120-$275	
		112 $180-$425			Brass Joints & Trim
	91 $140-$325	222 $225-$480	91 $220-$480	91 $125-$300	
		122 $200-$450			Brass Joints & Trim
98½ $150-$340	39 $140-$325	C221 $275-$600	39 $220-$500	39 $125-$300	
		C121 $260-$575			Brass Joints & Trim
91½ $140-$325		204 $225-$500			
		104 $200-$475			Brass Joints & Trim
92 $140-$325	92½ $125-$300	214 $225-$500	91½ $220-$475	91½ $120-$275	
		114 $200-$475			Brass Joints & Trim
99¼ $180-$425	40 $160-$375	C214 $300-$650	40 $250-$600	40 $150-$350	
		C114 $280-$625			Brass Joints & Trim
		224 $250-$550			
		124 $240-$525			Brass Joints & Trim

1 FOOT, 4 FOLD IVORY RULES (continued)	A. Stanley 1854-1857	Stanley Rule&Level 1859-....	Stearns (early) 1853-1856	Stearns (late) 1856-1902	Chapin (early) 1835-185?	Chapin (late) 185?-1901
Square Joint, Full Bound, Broad, Caliper Slide			**51B** $525-$850	**54B** $500-$800		**79** $200-$450
Arch Joint, Middle Plates, Narrow						
Arch Joint, Middle Plates						
Arch Joint, Middle Plates, Broad						
Arch Joint, Middle Plates, Broad, Caliper Slide						
Arch Joint, Edge Plates, Narrow						
Arch Joint, Edge Plates	**88½** $1000-$1400 / **88½** $1000-$1400	**88½** $250-$450 / **88½** $250-$450			**88** $225-$450	**56** $175-$400
Arch Joint, Edge Plates, Broad				**51** $400-$750		
Arch Joint, Edge Plates, Broad, Caliper Slide				**53** $500-$850		
Arch Joint, Full Bound, Narrow						**57** $180-$425
Arch Joint, Full Bound	**88** $1000-$1400 / **88** $1000-$1400	**88** $250-$450 / **88** $250-$450			**89** $250-$475	**57** $180-$425
Arch Joint, Full Bound, Caliper Slide						**78** $200-$475

Stephens 1854-1901	Chapin-Stephens 1901-1929	Belcher Bros. 1822-1877	Standard Rule 1872-1889	Upson Nut 1889-1922	Lufkin 1924-....	
	39½ $140-$375	C224 $325-$700				
		C124 $300-$675				Brass Joints & Trim
			113 $550-$1000			Ivory/Ebony/Ivory Leg Construction
		205 $240-$525				
		105 $225-$500				Brass Joints & Trim
		215 $240-$525				
		115 $225-$500				Brass Joints & Trim
		225 $240-$525				
		125 $225-$500				Brass Joints & Trim
		C225 $350-$750				
		C125 $340-$725				Brass Joints & Trim
		206 $240-$525				
		106 $225-$500				Brass Joints & Trim
93 $175-$375	88½ $160-$350	216 $240-$525	88½ $240-$550	88½ $140-$325		
		116 $225-$500				Brass Joints & Trim
		226 $275-$575				
		126 $260-$550				Brass Joints & Trim
		208 $275-$575				
		108 $260-$550				Brass Joints & Trim
94 $175-$400	88 $160-$375	218 $275-$575	88 $240-$575	88 $150-$350		
		118 $260-$550				Brass Joints & Trim
99½ $200-$450	99¾ $175-$400	C218 $300-$650				
		C118 $280-$625				Brass Joints & Trim

1 FOOT, 4 FOLD IVORY RULES (continued)	A. Stanley 1854-1857	Stanley Rule&Level 1859-....	Stearns (early) 1853-1856	Stearns (late) 1856-1902	Chapin (early) 1835-185?	Chapin (late) 185?-1901
Arch Joint, Full Bound, Broad				**51B** $525-$850		
Arch Joint, Full Bound, Broad, Caliper Slide				**53B** $550-$900		**78** $220-$500
Double Arch Joint, Middle Plates, Narrow						
Double Arch Joint, Middle Plates						
Double Arch Joint, Middle Plates, Broad						
Double Arch Joint, Edge Plates, Narrow						
Double Arch Joint, Edge Plates						
Double Arch Joint, Edge Plates, Broad						
Double Arch Joint, Full Bound, Narrow						
Double Arch Joint, Full Bound						
Double Arch Joint, Full Bound, Broad						

Stephens 1854-1901	Chapin-Stephens 1901-1929	Belcher Bros. 1822-1877	Standard Rule 1872-1889	Upson Nut 1889-1922	Lufkin 1924-....	
		228 $290-$625				
		128 $275-$600				Brass Joints & Trim
	99¼ $180-$425	**C228** $325-$700				
		C128 $320-$675				Brass Joints & Trim
		209 $290-$625				
		109 $275-$600				Brass Joints & Trim
		219 $290-$625				
		119 $275-$600				Brass Joints & Trim
		229 $320-$675				
		129 $300-$650				Brass Joints & Trim
		E209 $290-$625				
		E109 $275-$600				Brass Joints & Trim
		E219 $290-$625				
		E119 $275-$600				Brass Joints & Trim
		E229 $320-$675				
		E129 $300-$650				Brass Joints & Trim
		B209 $320-$675				
		B109 $300-$650				Brass Joints & Trim
		B219 $320-$675				
		B119 $300-$650				Brass Joints & Trim
		B229 $375-$800				
		B129 $360-$775				Brass Joints & Trim

2 FOOT, 2 FOLD IVORY RULES	A. Stanley 1854-1857	Stanley Rule&Level 1859-....	Stearns (early) 1853-1856	Stearns (late) 1856-1902	Chapin (early) 1835-185?	Chapin (late) 185?-1901
Square Joint, Gunter's Slide			46 $1700-$3000	46 $1700-$3000		
Square Joint, Full Bound, Gunter's Slide			46B(*) $2000-$3500	46B $2000-$3500		
Arch Joint					98 $1600-$2400	
Arch Joint, Gunter's Slide			45 $1800-$3200	45 $1800-$3200		
Arch Joint, Full Bound					99 $1800-$2800	
Arch Joint, Full Bound, Gunter's Slide			45B(*) $2500-$4000	45B $2500-$4000		

2 FOOT, 4 FOLD IVORY RULES	A. Stanley 1854-1857	Stanley Rule&Level 1859-....	Stearns (early) 1853-1856	Stearns (late) 1856-1902	Chapin (early) 1835-185?	Chapin (late) 185?-1901
Round Joint, Middle Plates, Narrow						
Round Joint, Middle Plates						
Round Joint, Middle Plates, Broad						
Square Joint, Middle Plates, Narrow						
Square Joint, Middle Plates						
Square Joint, Middle Plates, Broad						
Square Joint, Edge Plates, Narrow	85 $1000-$1600	85 $300-$600				59 $300-$70
	85 $1000-$1600	85 $400-$800				
Square Joint, Edge Plates			50 $625-$1000	50 $625-$1000	85 $400-$750	

Stephens 1854-1901	Chapin-Stephens 1901-1929	Belcher Bros. 1822-1877	Standard Rule 1872-1889	Upson Nut 1889-1922	Lufkin 1924-....	
						Engineers' Sliding Rule
						Engineers' Sliding Rule
						Engineers' Sliding Rule
						Engineers' Sliding Rule

Stephens 1854-1901	Chapin-Stephens 1901-1929	Belcher Bros. 1822-1877	Standard Rule 1872-1889	Upson Nut 1889-1922	Lufkin 1924-....	
		240 $275-$600				
		140 $280-$575				Brass Joints & Trim
		260 $275-$600				
		160 $280-$575				Brass Joints & Trim
		270 $325-$700				
		170 $325-$675				Brass Joints & Trim
		241 $350-$750				
		141 $340-$725				Brass Joints & Trim
		261 $350-$750				
		161 $340-$725				Brass Joints & Trim
		271 $390-$850				
		171 $375-$800				Brass Joints & Trim
77 $280-$675	85 $250-$600	242 $325-$750	85 $500-$950	85 $220-$500		
		142 $340-$725				Brass Joints & Trim
		262 $350-$750				
		162 $340-$725				Brass Joints & Trim

2 FOOT, 4 FOLD IVORY RULES (continued)	A. Stanley 1854-1857	Stanley Rule&Level 1859-....	Stearns (early) 1853-1856	Stearns (late) 1856-1902	Chapin (early) 1835-185?	Chapin (late) 185?-1901
Square Joint, Edge Plates, Broad			48 $750-$1200	49 $700-$1100	93 $350-$800	
Square Joint, Full Bound, Narrow						
Square Joint, Full Bound			50B(*) $750-$1200	50B $750-$1200		
Square Joint, Full Bound, Broad			48B(*) $800-$1500	49B $800-$1400		
Arch Joint, Middle Plates, Narrow						
Arch Joint, Middle Plates						
Arch Joint, Middle Plates, Broad						
Arch Joint, Edge Plates, Narrow				56 $750-$1200		
						59½ $500-$900
Arch Joint Edge Plates	86 $1000-$1800	86 $350-$700	49 $750-$1300	48 $750-$1200	86 $625-$1000	60 $550-$950
		86½ $450-$900				60½ $650-$1100
	86 $1400-$1800	86 $500-$1000				
Arch Joint, Edge Plates, Broad	94 $1500-$2000	94 $600-$1200	47 $750-$1300	47 $750-$1200	94 $625-$1000	63 $550-$950
	94 $1500-$2000	94 $600-$1200				
Arch Joint, Full Bound, Narrow				56B $750-$1300		

Stephens 1854-1901	Chapin-Stephens 1901-1929	Belcher Bros. 1822-1877	Standard Rule 1872-1889	Upson Nut 1889-1922	Lufkin 1924-....	
		272 $375-$850				
		172 $350-$800				Brass Joints & Trim
78 $500-$950		244 $420-$875				
		144 $375-$825				Brass Joints & Trim
		264 $420-$875				
		164 $375-$825				Brass Joints & Trim
		274 $460-$975				
		174 $425-$925				Brass Joints & Trim
		245 $450-$950				
		145 $425-$900				Brass Joints & Trim
		265 $450-$950				
		165 $425-$900				Brass Joints & Trim
		275 $500-$1100				
		175 $450-$1000				Brass Joints & Trim
		146 $425-$900				Brass Joints & Trim
	85½ $325-$775					Inside Beveled Edges
		246 $450-$950				
83 $500-$900	86 $350-$800	266 $450-$950	86 $900-$1300	86 $300-$675		
	86½ $500-$925					Inside Beveled Edges
		166 $425-$900				Brass Joints & Trim
	94½ $350-$800	276 $500-$1100				
		176 $450-$1000				Brass Joints & Trim
		248 $500-$1100				
		148 $450-$1000				Brass Joints & Trim

2 FOOT, 4 FOLD IVORY RULES (continued)	A. Stanley 1854-1857	Stanley Rule&Level 1859-....	Stearns (early) 1853-1856	Stearns (late) 1856-1902	Chapin (early) 1835-185?	Chapin (late) 185?-1901
Arch Joint, Full Bound	**87** $1600-$2200	**87** $350-$700	**49B(*)** $950-$1400	**48B** $900-$1300	**87** $500-$1000	**61** $400-$950
	87 $1700-$2400	**87** $600-$1200				
Arch Joint, Full Bound, Broad	**95** $1700-$2300	**95** $450-$900	**47B(*)** $1000-$1500	**47B** $1000-$1500	**95** $500-$1100	**64** $450-$1000
	95 $1800-$2500	**95** $600-$1200				
Arch Joint, Full Bound, Broad, Plain Slide						**65** $650-$1200
Arch Joint, Full Bound, Broad, Caliper Slide						**66** $650-$1200
Double Arch Joint, Middle Plates, Narrow						
Double Arch Joint, Middle Plates						
Double Arch Joint, Middle Plates, Broad					**96** $850-$1200	
Double Arch Joint, Edge Plates, Narrow						
Double Arch Joint, Edge Plates						
Double Arch Joint, Edge Plates, Broad	**96** $1000-$2500					
	96 $1000-$2500					
Double Arch Joint, Full Bound, Narrow						
Double Arch Joint, Full Bound	**89** $1400-$2000	**89** $600-$1000				**62** $600-$1100
	89 $2000-$2600	**89** $1000-$2200				
Double Arch Joint, Full Bound, Broad	**97** $1700-$2300	**97** $900-$1600			**97** $900-$1300	**67** $600-$1100
	97 $2000-$3000	**97** $1600-$2600				

Stephens 1854-1901	Chapin-Stephens 1901-1929	Belcher Bros. 1822-1877	Standard Rule 1872-1889	Upson Nut 1889-1922	Lufkin 1924-....	
84 $500-$900	**87** $350-$800	**268** $500-$1100	**87** $600-$1300	**87** $325-$750		
		168 $450-$1000				Brass Joints & Trim
87 $525-$950	**95** $350-$850	**278** $550-$1200	**95** $600-$1350	**95** $350-$800		
		178 $500-$1100				Brass Joints & Trim
	95½ $550-$1000					
	95¾ $550-$1000					
		249 $550-$1200				
		149 $500-$1100				Brass Joints & Trim
		269 $550-$1200				
		169 $500-$1100				Brass Joints & Trim
		279 $600-$1300				
		179 $550-$1200				Brass Joints & Trim
		E249 $550-$1200				
		E149 $525-$1100				Brass Joints & Trim
		E269 $550-$1200				
		E169 $500-$1100				Brass Joints & Trim
		E279 $600-$1300				
		E179 $550-$1200				Brass Joints & Trim
		B249 $650-$1400				
		B149 $600-$1300				Brass Joints & Trim
86 $600-$1000	**89** $525-$925	**B269** $650-$1400	**89** $600-$1500	**89** $525-$950		
		B169 $600-$1300				Brass Joints & Trim
88 $650-$1150	**97** $550-$1000	**B279** $750-$1600	**97** $700-$1600	**97** $650-$1100		
		B179 $700-$1500				Brass Joints & Trim

MISCELLANEOUS IVORY FOLDING RULES	A. Stanley 1854-1857	Stanley Rule&Level 1859-....	Stearns (early) 1853-1856	Stearns (late) 1856-1902	Chapin (early) 1835-185?	Chapin (late) 185?-1901
2 Foot, 3 Fold, Edge Plates,				**(No #)** $3000-$5000		
2 Foot, 3 Fold, Double Arch Joint, Narrow				**(No #)** $3200-$6000		
2 Foot, 6 Fold, Arch Joint, Edge Plates,				**50½** $1000-$1500		
				60 $1000-$1500		
2 Foot, 6 Fold, Arch Joint, Full Bound,				**60B** $1100-$2000		

48

Stephens 1854-1901	Chapin-Stephens 1901-1929	Belcher Bros. 1822-1877	Standard Rule 1872-1889	Upson Nut 1889-1922	Lufkin 1924-....	
			??? $2100-$4500			Masonic Rule
						Masonic Rule
34 $650-$1200		S246 $650-$1400				
		S146 $600-$1300				Brass Joints & Trim
35 $750-$1400						

49

BOARD, LOG, AND WOOD MEASURES	A. Stanley 1854-1857	Stanley Rule&Level 1859-....	Stearns (early) 1853-1856	Stearns (late) 1856-1902	Chapin (early) 1835-185?	Chapin (late) 185?-1901
Stick, Flat, 2 Foot, Board Measure						
Stick, Flat, 3 Foot, Board Measure		43 $500-$1000				90 $65-$200
		43½ $100-$200				90½ $65-$200
		49 $150-$300				90¾ $65-$200
Stick, Flat, 3 Foot, Log Measure						91 $75-$225
						91½ $75-$225
Stick, Square, 2 Foot, Board Measure	44 $1300-$1900				44 $75-$225	
		46½ $150-$250				87s $50-$175
			16 $350-$750	65 $325-$700		
			15 $300-$650	66 $275-$600		
Stick, Square, 3 Foot, Board Measure	45 $1300-$1900				45 $100-$250	
		47½ $200-$400				88s $65-$200
				67 $325-$700		
Stick, Square, 4 Foot, Board Measure				68 $325-$700		
Stick, Octagonal, 2 Foot, Board Measure	46 $1200-$1700	46 $200-$400				
			18 $325-$700	69 $300-$650		86 $50-$175
			17 $300-$650	70 $275-$600		
					46 $75-$225	87o $50-$175

Stephens 1854-1901	Chapin-Stephens 1901-1929	Belcher Bros. 1822-1877	Standard Rule 1872-1889	Upson Nut 1889-1922	Lufkin 1924-....	
		5912 $75-$225				
		5913 $90-$240				Brass Head
	43¼ $50-$175					Brass Head, Hickory
	43½ $50-$175	5916 $100-$250		43½ $35-$140		Brass Head
	49 $50-$175	5917 $100-$260		49 $35-$140		Steel Head
		5915 $100-$250				Brass Head
	48¾ $65-$200					T Head
	48¼ $65-$200				NOT	T Head, Hickory
					LISTED	
	46½ $40-$150	928 $85-$225			(see	Brass Head
		926 $80-$210			notes)	
		925 $75-$200				
	47½ $50-$175					
	46 $50-$160	921 $90-$225		46 $25-$125		Flange End
		920 $75-$200				
	45¾ $50-$175					

51

BOARD, LOG, AND WOOD MEASURES (continued)	A. Stanley 1854-1857	Stanley Rule&Level 1859-....	Stearns (early) 1853-1856	Stearns (late) 1856-1902	Chapin (early) 1835-185?	Chapin (late) 185?-1901
Stick, Octagonal, 3 Foot, Board Measure	47 $1300-$1800	47 $200-$400				88o $65-$200
				67½ $325-$700		
Walking Cane, Octagonal, 3 Foot, Board Measure		47½ $300-$600				
	48 $1500-$1900	48 $500-$1000				89 $160-$375
Walking Cane, Octagonal, 3 Foot, Log Measure		48½ $350-$700				89½ $160-$375
Wood Measure, 5 Foot	41½ $1600-$2000					

BENCH AND SHRINKAGE RULES	A. Stanley 1854-1857	Stanley Rule&Level 1859-....	Stearns (early) 1853-1856	Stearns (late) 1856-1902	Chapin (early) 1835-185?	Chapin (late) 185?-1901
Bench Rule, 1 Foot by 1¼ Inch						
Bench Rule, 1 Foot by 1 Inch, Brass Tips		34¼ $15-$40				
		34¼V $10-$30				
		34½ $25-$60				
		34½V $15-$30				
Bench Rule, 2 Foot by 1 Inch, Brass Tips						
Bench Rule, 2 Foot by 1¼ Inch						
Bench Rule, 2 Foot by 1¼ Inch, Brass Tips		31 $350-$700			31 $75-$200	
	34 $500-$1000	34 $15-$50	14 $200-$450	64 $175-$400		80 $40-$15#
		34V $10-$30				
			13 $350-$750	63 $325-$700		
Bench Rule, 2 Foot by 1¼ Inch, Bound				63½ $375-$800		

Stephens 1854-1901	Chapin-Stephens 1901-1929	Belcher Bros. 1822-1877	Standard Rule 1872-1889	Upson Nut 1889-1922	Lufkin 1924-....	
	47 $50-$175			47 $30-$125		
		922 $175-$400				
	48 $140-$325	923 $200-$450		48 $100-$240		Steel Head
		924 $225-$500				
	48½ $140-$325	929 $225-$500		48½ $100-$260	28½ $100-$240	

Stephens 1854-1901	Chapin-Stephens 1901-1929	Belcher Bros. 1822-1877	Standard Rule 1872-1889	Upson Nut 1889-1922	Lufkin 1924-....
					7031 $15-$50
					34¼ $15-$50
					34½ $25-$75
					7030 $15-$50
		910 $50-$125			
					7031 $20-$60
124 $40-$140					
	34 $25-$80			34 $20-$60	34 $20-$55
					7030 $20-$60

BENCH AND SHRINKAGE RULES (continued)	A. Stanley 1854-1857	Stanley Rule&Level 1859-....	Stearns (early) 1853-1856	Stearns (late) 1856-1902	Chapin (early) 1835-185?	Chapin (late) 185?-190?
Bench Rule, 2 Foot by 1½ Inch						
Bench Rule, 2 Foot by 1½ Inch, Brass Tips					30 $250-$500	
	35 $1400-$2000	35 $500-$1000				81 $200-$45
Bench Rule, 2 Foot by 1½ Inch, Brass Tips, Slide				62½ $550-$1200		
Bench Rule, 2 Foot by 1½ Inch, Bound						
Flat Rule, 3 Foot by 1¼ Inch						
Flat Rule, 3 Foot by 1¼ Inch, Brass Tips		71 $30-$70				
Flat Rule, 3 Foot by 1½ Inch, Brass Tips		71 $30-$70				
		80 $25-$150		80 $250-$550		
Flat Rule, 3 Foot by 1½ Inch, Bound						
Flat Rule, 4 Foot by 1¼ Inch, Brass Tips		71 $30-$70				
Flat Rule, 4 Foot by 1½ Inch, Brass Tips		71 $20-$50				
Flat Rule, 4 Foot by 1½ Inch, Brass Tips						
Flat Rule, 5 Foot by 1¼ Inch, Brass Tips		71 $30-$70				
Flat Rule, 5 Foot by 1½ Inch, Brass Tips		71 $30-$70				
Flat Rule, 5 Foot by 1½ Inch, Bound						
Flat Rule, 6 Foot by 1¼ Inch, Brass Tips		71 $40-$80				
Flat Rule, 6 Foot by 1½ Inch, Brass Tips		71 $40-$80				
Flat Rule, 6 Foot by 1½ Inch, Bound						
Shrinkage Rule, 2 Foot		30 $40-$90		81 $175-$450		
		30½ $20-$60				

Stephens 1854-1901	Chapin-Stephens 1901-1929	Belcher Bros. 1822-1877	Standard Rule 1872-1889	Upson Nut 1889-1922	Lufkin 1924-....
		911 $40-$100			
	35 $165-$380			35 $140-$320	
	35¼ $25-$100	915 $50-$125			
		912 $50-$125			7131 $20-$60
		914 $75-$175			7131½ $25-$120
					7031 $25-$75
					7030 $25-$75
					7132 $20-$60
	80 $25-$100				
		916 $55-$140			
					7132½ $25-$120
					7133 $25-$75
					7133½ $25-$130
					7134 $25-$75
					7134½ $25-$130
					7135 $25-$75
					7135½ $40-$150
135 $20-$75	30 $15-$65	??? $40-$100			820(x) $12-$50

Board Tables

BENCH AND SHRINKAGE RULES (continued)	A. Stanley 1854-1857	Stanley Rule&Level 1859-....	Stearns (early) 1853-1856	Stearns (late) 1856-1902	Chapin (early) 1835-185?	Chapin (late) 185?-1901
Shrinkage Rule, 2 Foot, 2 Fold		31 $250-$500		82 $275-$600		
		31½ $200-$400				
Extension Stick, 1½ to 3 Foot,		675 $300-$500				
Extension Stick, 2 to 4 Foot,		240 $40-$80				
Extension Stick, 2 to 4 Foot,						
Extension Stick, 2 to 6 Foot,						
Extension Stick, 3 to 6 Foot,		360 $40-$80				
Extension Stick, 4 to 8 Foot,		480 $50-$100				
Extension Stick, 5 to 10 Foot,		510 $50-$100				
		H510 $75-$150				
Extension Stick, 6 to 12 Foot,		612 $150-$300				

WANTAGE AND GAUGING RODS	A. Stanley 1854-1857	Stanley Rule&Level 1859-....	Stearns (early) 1853-1856	Stearns (late) 1856-1902	Chapin (early) 1835-185?	Chapin (late) 185?-1901
Wantage Rod, 16½ Inch		37 $250-$500				
		44 $200-$400			48 $150-$350	92 $125-$300
Gauging Rod, 3 Foot		45 $250-$500		78 $250-$550		93 $125-$300
Gauging Rod, 3½ Foot						
Gauging Rod, 4 Foot						94 $150-$350
					47 $175-$400	
Gauging Rod, Wantage Tables, 4 Foot		45½ $350-$700		79 $350-$750		94½ $150-$350
Gauging Rod, 4½ Foot						

Stephens 1854-1901	Chapin-Stephens 1901-1929	Belcher Bros. 1822-1877	Standard Rule 1872-1889	Upson Nut 1889-1922	Lufkin 1924-....	
37 $45-$150					820(x) $20-$100	
						Friction Clamp
					7162 $25-$125	Screw Clamp
		??? $75-$175				Friction Clamp
	98 $120-$225					Friction Clamp
					7172 $50-$180	Screw Clamp
					7163 $40-$140	Screw Clamp
					7164 $40-$140	Screw Clamp
					7165 $40-$150	Screw Clamp
						Screw Clamp, Folding Hook
					7166 $45-$160	Screw Clamp

Stephens 1854-1901	Chapin-Stephens 1901-1929	Belcher Bros. 1822-1877	Standard Rule 1872-1889	Upson Nut 1889-1922	Lufkin 1924-....	
	37 $120-$275	949 $200-$450		37 $90-$225	7188 $70-$210	12 Lines
	44 $100-$250	948 $175-$400		44 $90-$225	7187 $65-$200	8 Lines
		946 $175-$400				
		947 $175-$400				
	45 $100-$250	940 $150-$350		45 $90-$225	7181 $65-$200	120 Gallon Scale
		941 $175-$375				200 Gallon Scale
	45¼ $125-$300					
		943 $175-$375				300 Gallon Scale
	45½ $125-$300			45½ $100-$250		180 Gallon Scale
		944 $175-$400				300 Gallon Scale

YARD STICKS	A. Stanley 1854-1857	Stanley Rule&Level 1859-....	Stearns (early) 1853-1856	Stearns (late) 1856-1902	Chapin (early) 1835-185?	Chapin (late) 185?-1901
Yard Stick, ¾ Inch		33 $5-$30				97 $20-$75
Yard Stick, ¾ Inch, Brass Tips		50 $15-$50				98 $25-$100
Yard Stick, 1 Inch		211C $15-$50				
		214 $10-$50				
			19 $160-$325	72 $125-$275		82 $20-$75
						82½ $20-$75
Yard Stick, 1 Inch, Brass Tips	41 $500-$800	41 $5-$25	20 $175-$350	71 $140-$300	40 $45-$140	83 $25-$90
		214T $35-$70				
					41 $45-$140	
Yard Stick, 1¼ Inch, Brass Tips						
Yard Stick, ½ Inch Square, Brass Tips						97 $25-$125
Countertop Yard Measure, Flat, Metal		450 $500-$1000				
		550 $500-$1000				

OTHER MISCELLANEOUS RULES	A. Stanley 1854-1857	Stanley Rule&Level 1859-....	Stearns (early) 1853-1856	Stearns (late) 1856-1902	Chapin (early) 1835-185?	Chapin (late) 185?-1901
Rule, Wood, Length Unknown		174 $500-$1000				
Rule, Wood, 3 Inch, Cotton Staple Gauge		299 $450-$900				
Rule, Wood, 3 Inch, Caliper Slide						
		23 $600-$1200			49 $100-$250	

Stephens 1854-1901	Chapin-Stephens 1901-1929	Belcher Bros. 1822-1877	Standard Rule 1872-1889	Upson Nut 1889-1922	Lufkin 1924-....	
				33 $15-$50		Hickory
112 $25-$95	50 $25-$85			50 $20-$70	7422 $20-$65	Hickory, Rounded Edges
	33½ $20-$65	930 $25-$75			7121 $15-$50	
	33 $20-$65	931 $25-$75				
111 $20-$85	41 $20-$75	932 $35-$90	41 $35-$120	41 $20-$65	7122 $15-$60	
						Metal Edge, Eyelet
		933 $50-$125				Satinwood
		934 $50-$125				Satinwood
		935 $60-$140				Satinwood, Extra Thick
					7116 $12-$40	
					7129 $12-$40	
	41½ $25-$100					
					1068 $35-$125	
						Stapled to Wood Stick
					1069 $35-$125	Also Has Inch Scale

Stephens 1854-1901	Chapin-Stephens 1901-1929	Belcher Bros. 1822-1877	Standard Rule 1872-1889	Upson Nut 1889-1922	Lufkin 1924-....	
						Drafting, w/12 Scales
39 $30-$100						Architects' Rule
						Metric & English Graduations
				100 $25-$110	013 $25-$100	
						Button Gauge

OTHER MISCELLANEOUS RULES (continued)	A. Stanley 1854-1857	Stanley Rule&Level 1859-....	Stearns (early) 1853-1856	Stearns (late) 1856-1902	Chapin (early) 1835-185?	Chapin (late) 185?-1901
Rule, Wood, 3 Inch, Caliper Slide (continued)		**210** $800-$1400				
Rule, Ivory, 3 Inch, Caliper Slide		**24** $1000-$2000				
Rule, Wood, 4 Inch, Caliper Slide		**136** $10-$30				
Rule, Wood, 5 Inch, Plain Slide, w/Tables		**212** $125-$275				
Rule, Wood, 6 Inch, Caliper Slide		**136½** $10-$30				
Rule, Wood, 6 Inch, End Hook						
Rule, Wood, 1 Foot, Flat	**49** $1300-$1800					
		172 $500-$1000				
	50 $1000-$1400					**50** $20-$8.
		170 $300-$550				
		98 $75-$175				**99** $30-$10(
		99 $125-$300				**99½** $30-$10(
		23 $175-$400				
		98M $100-$250				
		34¼ $35-$90				
		34½ $50-$125				
					39 $175-$400	

Stephens 1854-1901	Chapin-Stephens 1901-1929	Belcher Bros. 1822-1877	Standard Rule 1872-1889	Upson Nut 1889-1922	Lufkin 1924-....	
						Button Gauge
						Button Gauge
						Inside-Measuring
					014 $30-$100	
					024 $30-$125	Extra Wide
					024B $35-$175	Extra Wide, Button Gauge
						Hatters' Rule
						Inside-Measuring
					016 $30-$125	
					026 $45-$150	Extra Wide
					046 $100-$250	Extra Wide, Spoke Caliper
					047 $50-$175	Coopers' Hook Stave Rule
						Drafting
						Drafting, 2 Scales
		No # $40-$100				Scholars' Rule
						Scholars' Rule
		No # $50-$125				School/Desk Rule, Beveled Edge
						School Rule, Beveled Edges
						Scholars' Rule, Beveled Edge
						Desk Rule, Beveled Edges
						School Rule, Maple
						School Rule, Boxwood
						"Platting" Scale

OTHER MISCELLANEOUS RULES (continued)	A. Stanley 1854-1857	Stanley Rule&Level 1859-....	Stearns (early) 1853-1856	Stearns (late) 1856-1902	Chapin (early) 1835-185?	Chapin (late) 185?-1901
Rule, Wood, 1 Foot, Flat (Continued)						
Rule, Wood, 1 Foot, Square,		**186** $600-$1200				
Rule, Brass, 1 Foot, End Hook						
Bevel, Wood, 1 Foot, Ship Carpenters', Sgl Blade	**42** $1200-$1800	**42** $400-$800 **43** $350-$700		**73** $275-$600 **74** $275-$600		**85½** $30-$125
Bevel, Wood, 1 Foot, Ship Carpenters', Dbl Blade	**42** $600-$1000	**42** $75-$150		**73** $160-$375 **77** $160-$375	**42** $60-$150	**85** $25-$100
Bevel, Wood,14 Inch, Ship Carpenters', Sgl Blade	**42½** $1000-$1500					
Bevel, Wood,16 Inch, Ship Carpenters', Sgl Blade						
Rule, Wood, 2 Foot, Flat		**176** $1000-$1600			**38** $175-$400	
Rule, Wood, 2 Foot, Flat, Half Bound						
Rule, Wood, 2 Foot, Flat, Full Bound						
Rule, Brass,2 Foot,2 Fold						
Rule, Brass,2 Foot,2 Fold, Zig-Zag		**17** $40-$80				

Stephens 1854-1901	Chapin-Stephens 1901-1929	Belcher Bros. 1822-1877	Standard Rule 1872-1889	Upson Nut 1889-1922	Lufkin 1924-....	
		951 $225-$500				Gunter's Scale
		950 $250-$550				Gunter's Scale, Satinwood
						Printers' Rule
					1063 $30-$100	Blacksmiths' Rule
30 $35-$120	43 $30-$100		43 $50-$170	43 $25-$90		
		900 $75-$180				Ungraduated
31 $25-$95	42 $20-$85	902 $40-$100	42 $40-$135	42 $20-$70	42 $20-$65	
		901 $40-$100				Ungraduated
						Rosewood, Ungraduated
		903 $75-$200				
		904 $85-$225				Thumb Screw
		956 $75-$180				Gunter's Scale
		952 $85-$220				Gunter's Scale, Satinwood
						"Platting" Scale
		957 $100-$240				Gunter's Scale
		953 $100-$250				Gunter's Scale, Satinwood
		958 $125-$300				Gunter's Scale
		954 $150-$340				Gunter's Scale, Satinwood
					1085 $35-$120	Blacksmiths' Rule
					1086 $40-$130	Blacksmiths' Rule, Circumference Scale
						Blacksmiths' Rule

OTHER MISCELLANEOUS RULES (continued)	A. Stanley 1854–1857	Stanley Rule&Level 1859–....	Stearns (early) 1853–1856	Stearns (late) 1856–1902	Chapin (early) 1835–185?	Chapin (late) 185?–1901
Rule, Wood, 1 Meter, Flat		**142** $60–$150				
		142M $60–$150				
Rule, Wood, 1 Meter, Flat, Brass Tips		**141** $60–$150				
		141M $60–$150				
Stick, Wood, 4 Foot, End Hook						
Stick, Wood, 5 Foot, T Head		**49½** $600–$1200				
Stick, Wood, 5 Foot, End Hook						
Stick, Wood, 4 Foot, Caliper						
Stick, Wood, 6 Foot, End Hook						

Stephens 1854-1901	Chapin-Stephens 1901-1929	Belcher Bros. 1822-1877	Standard Rule 1872-1889	Upson Nut 1889-1922	Lufkin 1924-....	
					7111ME $20-$75	Metric & English Graduations
					7111MM $20-$75	Metric Graduations
					7112ME $20-$75	Metric & English Graduations
					7112MM $20-$75	Metric Graduations
					7154 $40-$140	Freight Rule
						"Forwarding" Stick
		??? $100-$250			**7155** $40-$140	Freight Rule
		??? $175-$400				Freight Rule
					7156 $40-$140	Freight Rule

www.ingramcontent.com/pod-product-compliance
Lightning Source LLC
Chambersburg PA
CBHW020757220326
41597CB00012BA/567